This book belongs to:

....

....

To my snoozy dog, Holly – B.L.
To my sister Rosa, with love – M.L.

A TEMPLAR BOOK

First published in the UK in 2024 by Templar Books,
an imprint of Bonnier Books UK
4th Floor, Victoria House
Bloomsbury Square, London WC1B 4DA
Owned by Bonnier Books
Sveavägen 56, Stockholm, Sweden
www.bonnierbooks.co.uk

1 3 5 7 9 10 8 6 4 2

ISBN 978-1-80078-503-8

Edited by Ruth Symons
Designed by Chris Stanley
Production by Ché Creasey

Printed in China

FSC
www.fsc.org
MIX
Paper | Supporting responsible forestry
FSC® C104723

TIME FOR BED,
Animals
Ben Lerwill
Maribel Lechuga

templar
books

SLEEP

Sleep is amazing. When you go to sleep at night, it keeps you happy and healthy.

It makes your brain and body ready for another busy day. It even helps you grow!

Animals need sleep too.
Just like you, they have
to get lots of rest.
But the things they do at bedtime are sometimes
very different to the things you do...

DOGS

Dogs like to lick their paws before they go to sleep. It helps them to feel relaxed and keeps their paws clean.

What do you do before you get into bed?

Before they lie down, dogs often turn in circles to find the perfect spot. They like feeling safe and comfortable.

Puppies need more sleep than adult dogs.
Very old dogs need lots of sleep too.

CHIMPANZEES

Chimpanzees build their own beds, using strong sticks and branches. They make their beds in trees.

What do you like best about your bed?

Chimpanzees live in big family groups called troops.

Sleeping in the treetops keeps them safe from any prowling predators

DOLPHINS
Dolphins sleep with one eye open! These clever animals are always half-awake, even in the middle of the night.
Do you feel nice and sleepy when you close your eyes at bedtime?
Amazingly, when dolphins sleep, only one half of their brain goes to sleep. During the night, they can swap which side of their brain is awake, and which side is resting.

By staying half-awake, dolphins can float up to the surface when they need to breathe.

CATS

Cats have dreams too. Sometimes we can even see their legs and ears twitching! What do you think cats dream about?

How about you? What's your favourite kind of dream?

Just like you, cats like snoozing in places that are warm and cosy.

Some cats snore! This happens when their bodies are very relaxed. Dogs, lions and bears sometimes snore too – and so do people!

SEA OTTERS

Sea otters hold paws with each other while they sleep. It stops them floating away from each other in the night.

Have you ever fallen asleep next to somebody else?

Sea otters have thicker fur than any other animal. This helps them stay warm in the water.

Mummy sea otters sometimes wrap their babies up in seaweed, to keep them safe while they find food.

ANTS

Ants don't have one long sleep. Instead they have lots of very short naps throughout the day and night.

Do you ever nap in the daytime?

Some ants have 250 one-minute naps in a single day!

Every ant colony has a queen. The queen ant gets much more sleep, sometimes resting for nine hours at a time.

KANGAROOS

Baby kangaroos, or joeys, sleep in their mother's pouches. This keeps them safe, warm and snuggly.

Have you ever seen a grown-up carrying a sleeping baby?

Grown-up kangaroos lie down to go to sleep, just like you.

In very hot weather, kangaroos sometimes dig shallow sleeping holes, to stay cool.

BATS

Bats sleep upside down! They use their long claws to hang onto the roof of a cave, or the branches of a tree.

Do you sleep on your back, on your side, or on your tummy?

Bats are only active at night.
They sleep right through the day!

Baby bats hold on tight to
their mummies. Mother bats can
even fly while carrying their babies.

BIRDS

Birds are very different to us. Some birds sleep while they're flying! They keep their wings stretched out to stay high in the air.

Would you feel cosy up in the sky or would you rather be tucked up in bed?

Not all birds sleep in the air. Other birds spend the night resting in trees, bushes or other sheltered places.

Some birds, like ducks and geese, tuck their beaks into their feathers when they sleep, to keep warm.

FISH

Fish can't shut their eyes when they sleep – because they don't have eyelids! Some fish are awake in the day and some are awake in the night.

Some rest by burrowing into the sand,
or sheltering between rocks.
Others just float in the water.

GIRAFFES

Giraffes can sleep standing up! This helps them to stay safe and alert if animals like lions or leopards come too close.

Even though they're big and tall, giraffes don't need as much sleep as you do. They usually have just a few short naps each day.

Some wildlife experts used to think that giraffes didn't sleep at all!

TORTOISES

Baby tortoises sometimes doze for up to 22 hours a day! They bring their heads into their shells and fall fast asleep.

Just like you, tortoises like sleeping in the same place every night.

The perfect place for a tortoise to rest is somewhere quiet, dark and not too cold.

GOODNIGHT

When you snuggle down at bedtime, it's time for you to rest.

Getting lots of sleep means you can have an exciting day tomorrow.

Goodnight animals!
And goodnight you...

Five TOP TIPS

for getting a good night's sleep

1. Have a relaxing bedtime routine before you go to bed. This can include having a bath, brushing your teeth, reading a book and talking about your day.

2. Make your bedroom calm, comfortable, and nice and dark. Use a nightlight if you like, but try to make sure screens and televisions stay in other rooms.

3. Take your favourite teddy or soft toy to bed with you.

4. Go to bed at roughly the same time every night. It's also a good idea to wake up at roughly the same time every morning!

5. Don't eat too much just before going to bed. Have a small healthy snack if you're hungry, but try to save bigger meals and sugary treats for earlier in the day.